Gilberto Araujo
Ágata Lages Gava
Frederico F.C.T. Freitas

Effects of lead treatment on renal function

Gilberto Araujo
Ágata Lages Gava
Frederico F.C.T. Freitas

Effects of lead treatment on renal function

Effects of Acute Administration of a Low Dose of Lead Acetate on Renal Function in Wistar Rats

ScienciaScripts

Imprint

Cover image: www.ingimage.com

This book is a translation from the original published under ISBN 978-613-9-65609-7.

Publisher:
Sciencia Scripts
is a trademark of
Dodo Books Indian Ocean Ltd. and OmniScriptum S.R.L publishing group

120 High Road, East Finchley, London, N2 9ED, United Kingdom
Str. Armeneasca 28/1, office 1, Chisinau MD-2012, Republic of Moldova, Europe
Printed at: see last page
ISBN: 978-620-7-79208-5

SUMMARY

DEDICATORY

I dedicate this work to everyone who made me worthy".

ACKNOWLEDGMENTS

I thank God for the Gift of Life, for being in charge and for being the great responsible for all that I am.

My family for their support and commitment to my education from the very beginning.

To my advisor Prof.ª . Dr. Agata Lages Gava for the opportunity, the dedicated effort, the trust, the patience, the friendship and for all the teachings she has passed on. To you, my gratitude and admiration for the person you are and for your example of dedication to research.

To Prof. Dr. Elisardo Corral Vasquez and Prof.ª . Drª Silvana Meyrelles for their trust, support and inspiration, I will carry the great example that is in the person of each of you into my life.

To my friends in the laboratory for their companionship, I thank them for their friendship and for their tireless help at all times, which was fundamental to the completion of this work.

The Aguiar family, especially Mrs. Elaine, for believing in people and teaching me the essence of helping.

The Oliveira family for their hospitality and honesty. Especially Mr. Sebastiâo, a true intellectual, and his son William, a great friend.

To UFES and the Biotechnology Postgraduate Program for the opportunity to take this course

To the support and funding agencies: FAPES, CAPES and CNPq (PRONEX).

To all the friends and other people who have not been mentioned, but who contributed in some way to the development of this work.

SUMMARY

Lead is one of the most toxic environmental contaminants for humans and other animals. Its harmful effects can affect the functioning of all organs and systems and depend on the plasma levels of this metal. Currently, a blood concentration of approximately 60µg/dL is considered safe. However, studies have shown that plasma values lower than this biological tolerance index have deleterious effects on the body. The aim of this study was therefore to assess the effects of acute administration of a low concentration of lead on kidney function. Three-month-old male Wistar rats were divided into two experimental groups: Control (CT, n=10) and Treated (TT, n=10). The animals were anesthetized and had their trachea, bladder, vein and femoral artery catheterized and received injections of 0.9% saline (CT) or lead acetate (TT, 320µg/kg *iv*) one hour before inulin and para-aminohippurate (PAH) *clearance.* Renal function was assessed by determining glomerular filtration rate (GFR), renal plasma flow (RPF), renal blood flow (RBF), filtration fraction (FF) and renal vascular resistance (RVR). After *clearance,* plasma was collected for plasma urea quantification. *p<0,05. The GFR (mL/min/Kg) in the animals in the TT group was increased (CT: 6.1 ± 0.1; TT: 7.1* ± 0.3) accompanied by a reduction in uremia (mg/dl) (CT: 46 ± 4; TT: 30* ± 5). FPR (mL/min/Kg) and FSR (mL/min) were increased in the TT group (23.01 ± 1.5 and 39.9 ± 2.9, respectively) when compared to the CT group (17.86 ± 1.63 and 29.5 ± 2.2, respectively). There was no change in FF (%) between the groups (CT: 38.7 ± 3%; TT: 34.8 ± 2%), but RVR (ua) was lower in the TT group compared to the CT group (CT: 3.4 ± 0.3; TT: 2.5* ± 0.1ua). We can conclude that lead treatment causes significant changes in kidney function, such as an increase in GFR and a reduction in uremia. There was also an increase in renal perfusion as evidenced by an increase in RPF and RBF, probably due to a reduction in RVR. These data demonstrate the toxic potential of lead on renal hemodynamics, even in the case of acute, low-dose exposure.

Keywords: Lead, Renal Function, Heavy Metals.

1. INTRODUCTION

1.1 Lead

Lead (Pb) is a naturally occurring element that has been widely used for thousands of years. Today, it is one of the most common contaminants in the environment, due to the numerous industrial activities that favor its widespread distribution. Due to its persistence in the environment, exposure to lead remains a serious public health problem worldwide (PATOCKA & CERNY, 2003). Thus, all human beings have lead in their bodies as a result of exposure to exogenous sources (SARYAN & ZENSZ, 1994), with ingestion and inhalation being the main forms of contamination, once absorbed by the gastrointestinal tract or nasal mucosa, it is distributed through the plasma accumulating in soft or mineralized tissues. Its toxic effects on humans and animals have been known for a long time as it affects practically every organ and system in the human body (XIE ET AL., 1998).

1.1.1 Form

Lead is a grayish-white, shiny, ductile and soft metal. It belongs to the chemical family of metals in group IVA of the periodic table, along with carbon, germanium, silicon and tin. It is found in nature both in its free state and in compositions with various other elements (PAOLIELLO & CHASIN, 2001).

Lead compounds are used as pigments in paints, dyes and ceramic glazes. The amount of lead used in these products has been reduced in recent years to minimize the harmful effect of lead on people and animals. Tetraethyl and tetramethyl lead were used in the United States as gasoline additives to increase its octane rating (BAIRD, 2002). In addition to these compounds, the interaction of lead with other elements gives rise to compounds such as lead sulphate, lead chromate, lead arsenate, lead dioxide, lead bromide, litargium (PbO), zirconium (Pb O_{34}) and alvaiade [PbCO3.Pb(OH)2] (SPiNOLA ET AL., 1980).

1.1.2 Occurrence

Lead is relatively abundant in the earth's crust, with an average concentration of between 10 and 20 mg/kg. The largest natural sources of lead are volcanic emissions, geochemical weathering and aquatic mists (WHO, 1995). It is estimated that natural lead emission rates are around 19,000 tons per year (WHO, 1995). However, anthropogenic processes are mainly responsible for the release of lead into the environment (ATSDR, 1993), with mining and smelting of primary lead (from galena ore) and secondary lead (from the recovery of scrap metal or batteries) being important sources of lead emissions (WHO, 1995). In addition, the combustion of gasoline containing lead is responsible for the presence of a large proportion of lead in the environment in countries that have not yet banned the use of this metal as an anti-knock agent in gasoline (BAIRD, 2002). Due to its physical characteristics of ductility, malleability and low melting point, lead has been used by ancient populations (MIDIO & MARTINS, 2000). Lead can be consumed as a metal, pure or alloyed with other metals, or as compounds in the form of oxides (ATSDR, 1993). Currently, its main use is in the production of batteries and in some types of solder, since it forms important alloys with other metals.

Lead is commonly found in soil, especially near roads, old houses, old orchards, mining areas, industrial facilities, near power plants, incinerators, landfills and hazardous waste sites. People living near hazardous waste sites can be exposed to lead and lead-containing chemicals by breathing the air, drinking water, eating food, or swallowing dust or dirt containing lead. People can be exposed to lead by ingesting food or drinking water that contains lead (GOYER AND CLARKON, 2001).

Young children have a constant habit of putting their hands and objects to their mouths, a behavior that increases the risk of exposure in the event of environmental contamination, as well as contact with the contaminated clothing of parents or adults who work with metal (CALABRESE, 1997; JIM, 1997).

Food can contain small amounts of lead. However, as lead solder is no longer used in cans, very little lead is found in food. Leafy vegetables grown in soil containing lead have lead dust on them. Lead can also get into food if it is improperly placed in glass, pottery or ceramic containers containing lead (ATSDR, 1999).

1.1.3 Toxicokinetics

Currently, lead poisoning in the human population occurs mainly through occupational exposure, within the hundreds of professions that use this element and its compounds (PAOLIELLO & CHASIN, 2001).

Non-occupational exposures in adults are restricted to those that occur through the ingestion of food and drink, representing the main source of daily absorption of inorganic compounds for the general population. Organic lead compounds are not found in food (MIDIO & MARTINS, 2000).

The absorption of lead in the body is influenced by factors such as the concentration and time of exposure to the metal, as well as factors related to the physicochemical properties such as particle size and solubility of the compound (SCHIFER ET AL., 2005).

Once absorbed, lead is distributed in the blood compartment in two forms: one non-diffusible, bound to erythrocytes, and the other diffusible, in the plasma. Although more than 95% of circulating metal remains bound to erythrocytes, the piasmatic form is the most active fraction of the metal. Due to its greater availability, this fraction can directly influence the concentration of lead in other compartments and, consequently, contribute to the metal's toxicity in different organs and systems (SILVA, 1983). The amount of lead in the blood, which generally represents short-term exposure, rises until it reaches a plateau, and the excess penetrates soft tissues such as the brain, liver and kidneys (MUSHAK & CROCETTI, 1989; BAIRD, 2002). The half-life of lead in the blood is 36 days and in soft tissues it is 40 days (PAOLIELO & CHASIN, 2001; KOSNETT, 2003). Lead is finally deposited in the bones, where it replaces calcium, since the ions Pb^2 + and Ca^2 + are similar in size (BAIRD, 2002). In adults, almost 90% of the body's lead load is contained in bone, where the half-life of this metal is 20 to 27 years (RABINOWITZ ET AL., 1976). The lead present in the skeleton is biologically inert, but in situations involving bone demineralization, such as pregnancy and osteoporosis, lead can be rapidly mobilized leading to the appearance of toxic symptoms, even after exposure has ceased (SARYAN & ZENZ, 1994). However, in conditions that do not involve the dissolution of bones, the toxicity of lead is proportional to the amount present in the soft tissues, and not to the amount found in the blood or bones (BAIRD, 2002). Lead is eliminated from the body through urine and feces (biliary excretion) and can also be excreted through breast milk. The concentration of lead secreted by breast milk varies between 10 and 30% of plumbemia (PAOLIELLO & CHASIN 2001).

1.2 Urinary System

The urinary system plays a part in maintaining homeostasis by eliminating metabolic waste, water and other substances through urine. This system is made up of the kidneys and the urinary tract. The urinary tract comprises the ureters, bladder and urethra. The urine produced in the kidneys is collected by the ureters and transported to the bladder, where it is temporarily stored. From the bladder, the urine is expelled into the external environment through a channel called the urethra.

The urinary system, in the form of the kidney, plays other roles in the body, such as: regulating water and electrolyte balance; excreting metabolic waste; excreting bioactive substances; regulating blood pressure; regulating the production of red blood cells; regulating vitamin D production; and gluconeogenesis (EATON, 2006).

1.2.1 Anatomy and Physiology of the Kidneys

The kidneys are a pair of organs that lie behind the peritoneal lining of the abdominal cavity. Each kidney is lined by a thin fibromuscular capsule. The typical adult kidney measures 10 to 12 cm in length, 5 to 7 cm in width and 3 cm in thickness, with its mass varying between 135 and 150g. The medial side of each kidney has an indented region known as the hilum, through which pass the renal artery and vein, lymphatic vessels, nerve supply and the ureter, which carries urine from the kidney to the bladder. In the bladder, urine is stored and periodically eliminated from the body. Each kidney is surrounded by a fibrous outer capsule that protects its internal structures. The longitudinal dissection of a kidney allows two distinct regions to be visualized: an outer region known as the cortex and an inner region known as the medulla. The medulla is subdivided into multiple tissue masses shaped like cones called renal pyramids. The base of each pyramid originates at the boundary between the cortical and medullary regions and ends in the papilla, which protrudes into the space of the renal pelvis, a funnel-shaped structure that continues with the upper end of the ureter. The outer edge of the pelvis is divided into blind-bottomed structures called major calyces, which are divided into minor calyces that collect urine from the tubules of each papilla. The walls of the calyces, pelvis and ureter contain contractile elements that propel the urine towards the bladder (GUYTON, 2006).

The kidney is a major organ in the body's homeostasis, both in terms of the number of functions it performs and the importance of these functions. Some of its responsibilities include regulating the hydroelectrolytic balance, the acid-base balance, as well as the excretion and reabsorption of metabolites produced in the body (CUNNINGHAM, 2004).

In mammals, the two kidneys receive approximately 25% of the cardiac output. The renal artery enters the kidney through the hilum and then progressively divides into interlobar arteries, arcuate arteries and interlobular arteries (also known as radial arteries) and afferent arterioles, which end in the glomerular capillaries, where large quantities of liquid and solute are filtered. The distal ends of the capillaries of each glomerulus coalesce to form the efferent arteriole, which gives rise to a second network of capillaries, the peritubular capillaries, which surround the renal tubules (GUYTON, 2006). The kidneys are prepared to detect excess or deficiency of water and specific electrolytes, and act by altering the rate of reabsorption or secretion of these substances. They also produce hormones that regulate systemic blood pressure and the production of red blood cells (CUNNHINGAM, 2004).

In order to carry out their functions, the kidneys have filtration, reabsorption and secretion processes, which take place in the nephrons, considered to be the smallest functional unit of the kidney. Each human kidney has approximately 1 million nephrons, which are not capable of regeneration (AIRES, 2008).

Each nephron has a vascular component, made up of the glomerulus and peritubular capillaries, and a tubular component, made up of the proximal convoluted tubule, loop of Henle, distal convoluted tubule and collecting duct. The glomeruli contain a network of capillaries which, compared to other capillaries in the body, have a high hydrostatic pressure (around 60 mmHg). The glomerular capillaries are covered by epithelial cells and every glomerulus is surrounded by Bowman's capsule. The liquid filtered by the glomerular capillaries then flows into Bowman's capsule and from there into the proximal convoluted tubule, which is located in the renal cortical area (GUYTON, 2006).

From the proximal convoluted tubule, the liquid flows into the loop of Henle, which plunges into the renal medulla. Each loop of Henle consists of a descending segment and an ascending segment. After the ascending segment, the liquid flows into the distal convoluted tubule, which, like the proximal convoluted tubule, is located in the cortical zone. The distal convoluted tubule is followed by the collecting duct, which join to form progressively larger ducts that empty into the renal pelvis through the papillae (GUYTON, 2006).

The formation of urine therefore begins in the glomerular capillaries, where around 20% of the plasma that enters the kidney is filtered out thanks to the high hydrostatic pressure in these capillaries. The glomerular filtrate is similar in composition to plasma. After its formation, the glomerular filtrate travels through the renal tubules, where it undergoes changes in its composition through the processes of reabsorption (passage of substances from the tubular lumen to the peritubular capillaries) and secretion (passage of substances from the peritubular capillaries to the tubular lumen) (AIRES, 2008). It is worth noting that the rates of reabsorption and secretion of the various substances vary in order to maintain their values at normal levels in the plasma. In order to maintain homeostasis, the excretion of water and electrolytes must be carefully matched to their respective gains. If the gain exceeds the excretion, or the excretion exceeds the intake, changes in the renal processes of reabsorption and secretion will be made in order to maintain normal plasma values of these substances (GUYTON, 2006). In this way, kidney function allows the body to retain and store the substances it needs, according to its needs, and to eliminate the substances it doesn't need and/or are in excess (J. ELLIOTT & D.A. ELLIOTT, 2009).

1.3 Lead and the Kidney

Lead affects various organs and systems in humans, with the renal system being one of the main targets (PAYTON *ET AL.*, 1994). This metal accumulates selectively in the kidneys (CHOIE & RICHT, 1980) and acute exposure affects the cells of the proximal tubule causing aminoaciduria, phosphaturia, glycosuria (GOYER, 1990) and decreased excretion of uric acid (PAOLIELLO & CHASIN, 2001). An inverse correlation has also been observed between creatinine clearance and blood lead levels in humans (STAESSEN *ET AL.*, 1992). Chronic exposure to this metal can cause renal tubular injury and interstitial nephritis, which can progress to chronic renal failure (GOYER, 1989).

The effects of lead on human health depend on the plasma levels of this metal and currently a blood concentration of approximately 60µg/dL is considered safe.

However, studies have shown that plasma levels lower than this biological tolerance index have deleterious effects on the cardiovascular system. Considering that the cardiorenal axis plays an extremely important role in body homeostasis, research into the possible toxic action of the metal at low doses is of great importance for understanding lead-related pathology.

2. OBJECTIVES

2.1 General Objective

To evaluate the acute effects of a low dose of lead acetate on the renal function of Wistar rats.

2.2 Specific Objectives

In rats treated with lead acetate and their respective controls, evaluate:

- Glomerular filtration rate;
- Renal plasma flow;
- Renal blood flow;
- Renal vascular resistance;
- Blood pressure;
- Heart rate;
- Plasma urea values;

3. MATERIALS AND METHODS

3.1 Animals and Experimental Groups

This study used 3-month-old male Wistar rats, supplied by the vivarium of the Postgraduate Program in Physiological Sciences at the Federal University of Espirito Santo. The animals were kept under controlled and stable conditions in a vivarium with a temperature of 22°C, humidity of 50-70% and a 12-hour light-dark cycle, where they had free access to water and food *ad libitum*. The experimental protocols for this project were approved by the Ethics Committee for the Use of Animals (CEUA) of the Federal University of Espirito Santo (UFES) under number 002/2012.

The animals were divided into two experimental groups according to the treatment they received: in the control group, the animals were treated with vehicle (0.9% saline) and the animals in the treated group were treated with lead acetate in a single intravenous injection at a dose of 320µg/kg, the choice of this dose was based on previous studies (SIMOES ET ALL., 2011) and the aim was to achieve a blood concentration of 32µg/dl.

3.2 Renal function assessment

The analysis of hemodynamics and renal function included the evaluation of the parameters of renal blood flow (RBF), renal plasma flow (RPF), renal vascular resistance (RVR), glomerular filtration rate (GFR) and filtration fraction (FF) through the determination of hematocrit and clearance of inulin and para-aminohippurate (PAH), as described by Alves et al. (2002).

To this end, the animals were anesthetized with sodium thiopental (50 mg/Kg i.p.)

and had their trachea catheterized (PE-240) to facilitate breathing. The animals' left femoral artery (PE 50), left femoral vein (PE 50) and bladder (PE 240) were catheterized for blood collection, inulin infusion, PAH and urine collection, respectively. Urine volume was determined gravimetrically (FERNANDEZ E MALNIC, 1995).

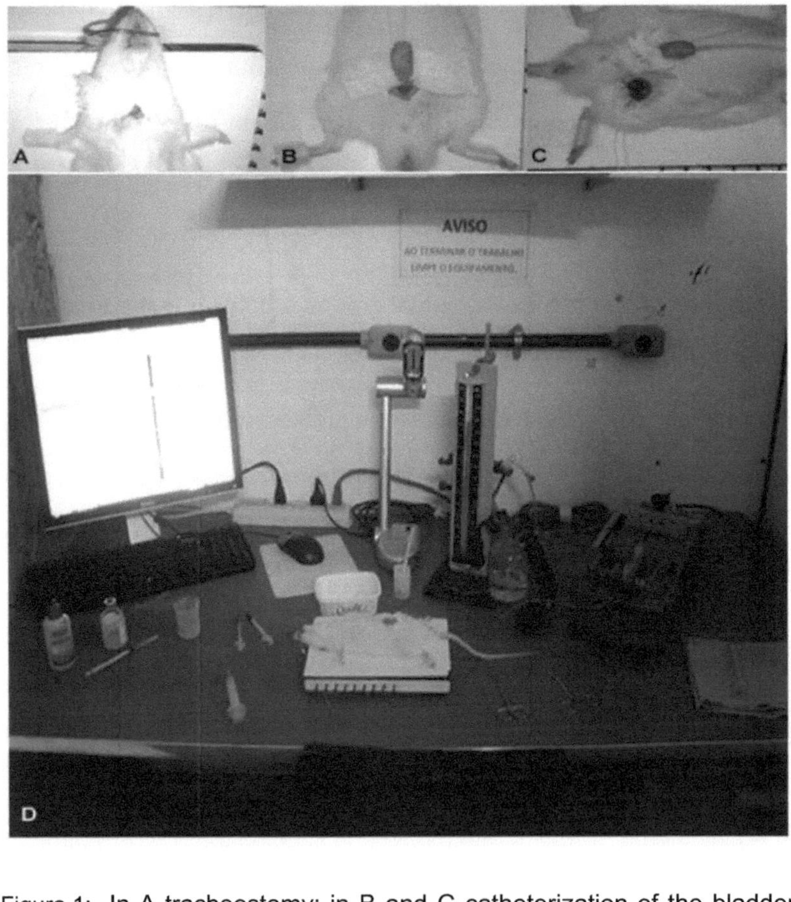

Figura 1: In A tracheostomy; in B and C catheterization of the bladder, artery and femoral vein, respectively. In D, the *setup of* the function with the animal positioned.

After the surgical procedures, the animals received an intravenous injection of lead acetate (320µg/kg) or vehicle (0.9% saline - 100µL).

After 1 hour, a solution of 3% mannitol was administered intravenously for 30 minutes in order to increase diuresis.

After these procedures, a prime dose of inulin and PAH was administered intravenously (30mg/100g of body weight of inulin and 0.02mg of PAH for every 1mL of extracellular volume (ECV), with ECV considered to be 20% of body weight) and after this procedure, infusion was started to maintain the concentration of these substances in the blood (0.015g/ml of inulin and 0.004g/ml of PAH). During the experiment, the animals were infused with an isotonic saline solution containing 3% mannitol at a rate of 0.1mL/min using a continuous infusion pump.

After a period of 30 minutes, four 30-minute urine collection periods were carried out. 5 minutes before the urine collections, blood was taken, as shown in figure 5.

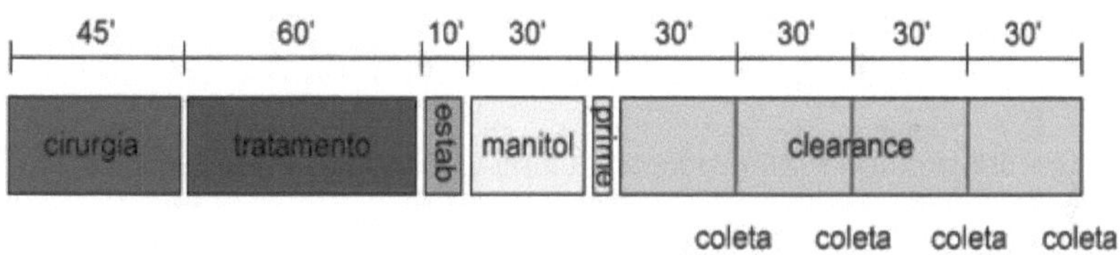

Figura 2: Chronological outline of the experimental protocol.

The plasma and urinary values of inulin were determined using the modified Fuehr method (1955) and the plasma and urinary values of PAH were quantified using the modified Smith method (1945).

3.2.1 Renal hemodynamics

Using urine flow and quantification of the plasma and urine values of inulin and

PAH, their clearance rates were calculated, which reflect the glomerular filtration rate and plasma flow of the animals, respectively. Insert clearance formula

The filtration fraction was determined by the ratio between the glomerular filtration rate and the renal plasma flow, according to the formula below.

$$FF = RFG/FPR$$

Blood samples were collected in capillary tubes and centrifuged for 5 minutes at 4000 rpm to determine the hematocrit.

Renal blood flow was calculated as renal plasma flow divided by 1- hematocrit, using the formula:

$$FSR = FPR/(1\text{-hematócrito})$$

During this time, the animals' mean arterial pressure was continuously recorded using a pressure transducer connected to a data acquisition system (BIOPAC MP100, Biopac Systems) and the data was analyzed using software.

The RVR was then determined mathematically using Ohm's law, where a:

$$RVR = PAM/FSR$$

3.2.2 Plasma Urea Values

The enzymatic colorimetric test (BIOCLIN, Belo Horizonte, Brazil) was used to determine plasma urea values. It is based on the principle that urea is hydrolyzed to ammonium ions and CO2 by the enzyme urease according to the equation below:

$$\text{Urea} + 3H_2O \xrightarrow{\text{Urease}} 2\,NH_4^+ + CO_2 + 2OH^-$$

At alkaline pH and in the presence of salicylate and sodium hypochlorite, ammonia produces a greenish compound whose color intensity is proportional to the concentration of urea in the sample analyzed.

The color intensity was quantified by reading it on a spectrophotometer (Analyser SP 220) using a wavelength of 600nm and the following formula was used to calculate the concentration of urea in the samples:

$$Urea\ (mg/dL) = \frac{\text{sample absorbance}}{\text{absorbing the standard}} \ x\ 70{,}0$$

3.3 Statistical Analysis

All data is expressed as mean ± SEM. For statistical analysis, *Student*'s t-test was used for independent samples and for measurements when necessary. Values were considered statistically significant when $^*p<0.05$ and $^{**}p<0.01$.

4. RESULTS

4.1 Renal Function Assessment

4.1.1 Glomerular filtration

Our first result was the assessment of inulin clearance. This technique is considered the gold standard for determining the Glomerular Filtration Rate (GFR) since inulin is freely filtered in the glomeruli without being reabsorbed or secreted. When we compared inulin clearance in the different groups, we found a significant increase in the TT group (7.1±0.3 mL/min/Kg) compared to the CT group (6.1±0.1 mL/min/Kg), as shown in figure 6.

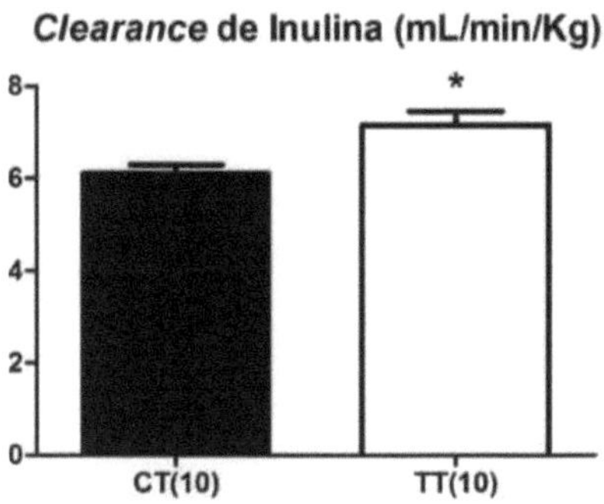

Figure 3 - Inulin clearance values in control (CT) and treated (TT) animals. Inulin clearance represents the Glomerular Filtration Rate. The number in brackets indicates the number of animals in each group. *p<0,05.

4.1.2 Renal Plasma Flow

Together with the assessment of Glomerular Filtration Rate, we also assessed Renal Piasmatic Flow by means of Para-aminohippurate Acid Clearance. As PAH is filtered in the glomerulus and secreted into the tubules without being reabsorbed, its clearance is an important marker of plasma flow through the renal capillaries.

When we compared the PAH clearance values found in the two groups, we observed a significant increase in the TT group (23.01±1.5 mL/min/Kg) compared to the CT group (17.86±1 mL/min/Kg) as shown in figure 4.

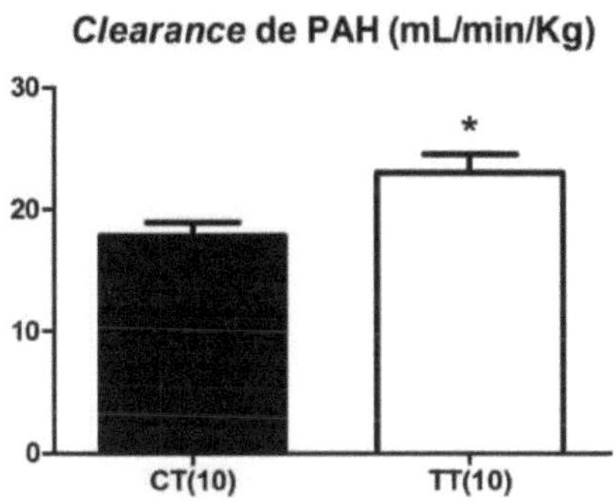

Figure 4 - PAH clearance values in control (CT) and treated (TT) animals. PAH clearance indicates Renal Plasma Flow. The number in brackets indicates the number of animals in each group. *p<0,05.

4.1.3 Renal Blood Flow

Renal blood flow can be obtained from the ratio between renal plasma flow and hematocrit during the renal clearance protocol. When we compared the RBF of the lead-treated animals, we observed a significant increase in the TT group (39.9± 2.9 ml/min/Kg) compared to the control CT group (29.5±2.2 mL/min/Kg), a result that can be seen in figure 5.

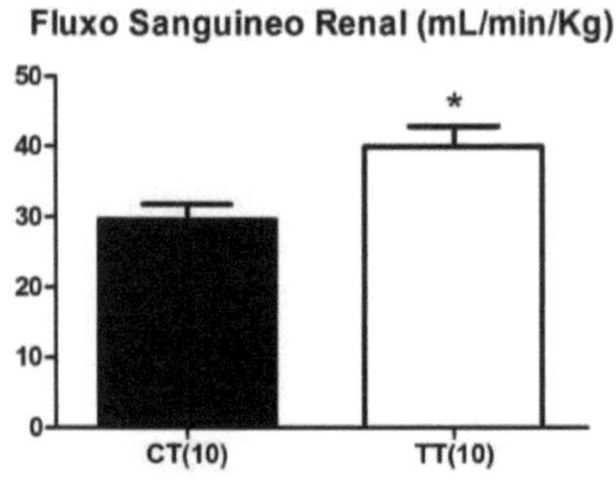

Figure 5 - Renal blood flow values in control (CT) and treated (TT) animals. The number in brackets indicates the number of animals in each group. *p<0,05.

4.1.4 Renal Vascular Resistance

The Renal Vascular Resistance obtained by Ohm's Law is an important parameter for analyzing renal hemodynamics. As can be seen in figure 9, the RVR of the TT group (2.5±0.1ua) was significantly lower than that of the CT group (3.4±0.3ua).

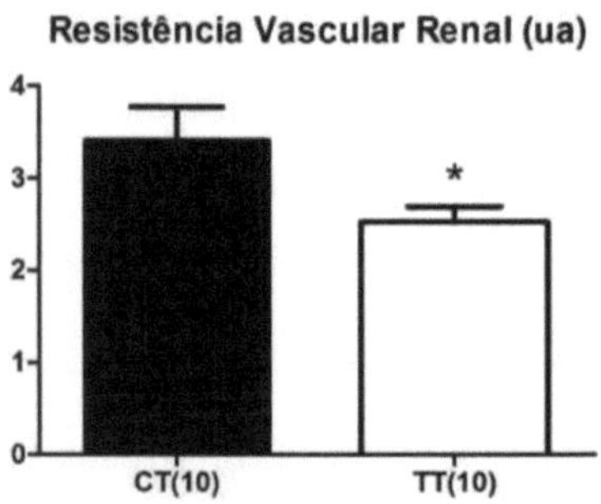

Figure 6 - Renal vascular resistance in control (CT) and treated (TT) animals. The number in brackets indicates the number of animals in each group. *p<0,05.

4.1.5 Filtration fraction

We also analyzed the filtration fraction (FF), which reflects the percentage of plasma volume that is filtered by the glomeruli. When comparing the FF data of the TT group (34.8±2%) with the CT group (38.7±3%), we observed no difference between the groups (Figure 10).

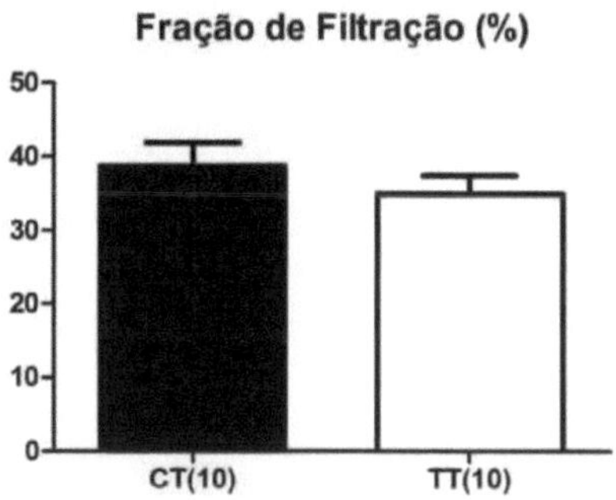

Figure 7 - Filtration Fraction (FF) values. The number in brackets indicates the number of animals in each group.

4.1.6 Piasmatic urea

Determining the concentration of pyosmotic urea is an important marker of kidney function. When comparing the TT group (30±5 mg/dL) with the CT group (46±4 mg/dL) we noticed a reduction in pyosmotic urea (figure 8).

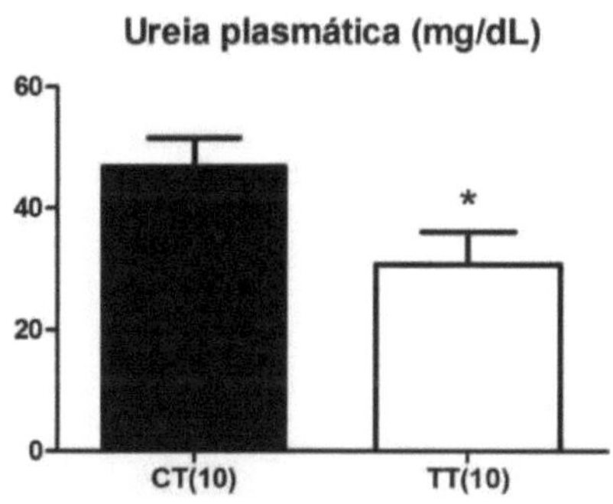

Figure 8 - Comparison of the pyosmotic urea concentration of the animals in the TT group in relation to the CT group. The number in brackets indicates the number of animals in each group. *p<0,05.

4.2 Cardiovascular Parameters

4.2.1 Heart Rate

The animals' heart rate and blood pressure were monitored throughout the treatment and clearance period. Figure 9 shows the heart rate of the animals in the TT group and the CT group.

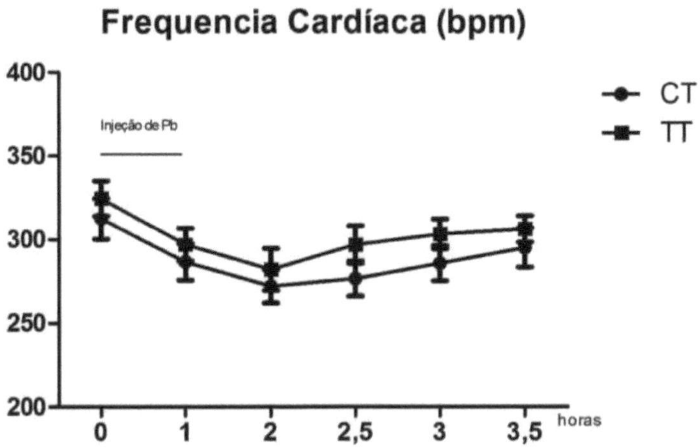

Figure 9 - Heart rate analysis of CT and TT animals. The 0-1 hour period was for treatment with Pb in the TT group. The other points correspond to the blood sampling periods during clearance.

4.2.2 Mean arterial pressure

Mean arterial pressure was monitored throughout the experiment in order to assess RVR (figure 10). Both groups showed a drop in MAP during the experimental protocol, which was statistically different only 2 hours after lead injection.

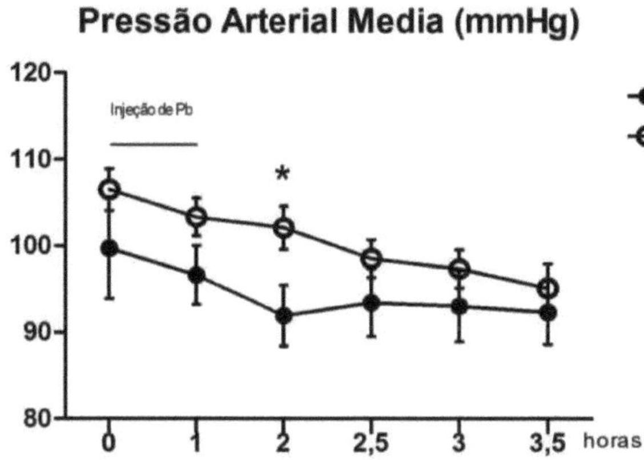

Figure 10 - Mean arterial pressure (MAP) of the CT and TT groups during treatment (0-1 hour) and clearance assessment (2-3.5 hours). The number in brackets indicates the number of animals in each group. *p<0,05.

5. DISCUSSION

This study shows significant changes in the renal hemodynamics of Wistar rats subjected to acute administration of a low dose of lead acetate. These alterations are evidenced by an increase in GFR, an increase in RPF, a reduction in RVR and a consequent increase in FSR. The aforementioned effects may be associated with changes in the dynamics of the renal arterioles, more specifically the relaxation of the afferent arteriole or the constriction of the efferent arteriole.

Several studies have shown that the toxic effect of lead on human health depends on blood concentration and exposure time (LUSTBERG, 2002). The Agency for Toxic Substances and Disease Registry (ATSDR) considers the reference value of 60µg⁄dl safe for the concentration of lead in the blood (PATRICK, 2006), and treatment with a lower concentration than the blood reference is relevant, as there are reports of damage to other systems when exposure to low concentrations of lead occurs.

Increased GFR may indicate an important renal response to the presence of toxic substances, such a response is known as the hyperfiltration hypothesis, and can be found in metal poisoning (WEAVER ET AL. 2003). Renal hyperfiltration has been frequently observed in various kidney diseases and clinical conditions such as type I diabetes, sickle cell disease, obesity, and high protein diet, but in most cases it is associated with clinical abnormalities such as hypertension (COURTOIS ET Al 2003, FRIEDMAN, 2004). Our results may indicate a renal hyperfiltration response to lead acetate intoxication, since renal excretion is one of the main elimination routes for the metal. According to Lyn Patrick (2006) the main means of elimination of this metal is the kidney and the increase in filtration rate as a response to renal toxicity. Renal hyperfiltration has already been described in foundry workers exposed to low concentrations of lead (ROELS ET AL, 1994; WEAVER ET AL, 2003).

In addition, the increase in renal plasma flow and, consequently, renal blood flow, found in our study is in line with the increase in glomerular filtration rate in renal hyperfiltration in an attempt to detoxify the body, demonstrating the harmful potential of lead at the dose presented. We can associate this result of increased flow with a reduction in renal vascular resistance. Corroborating our data, studies have shown that lead is capable of reducing the contraction capacity of smooth muscle (FIORIM, 2011). Changes in cytosolic Ca^2 + are closely involved in regulating vascular tone and vascular smooth muscle contraction. Several studies have revealed that lead can potentially compete with Ca^{2+} for transport through channels and pumps involved in the movement of ions in particular Ca^2 + across the cell membrane and between cytoplasm, endoplasmic reticulum and mitochondria (GOLDSTEIN, 1991 AND SIMONS, 1993). Lead may also be able to compete with calcium for binding sites or receptors, such as N-methyl aspartate (CORY-SLECHTA, 1996; MARCHIORO, 1996) and the nicotinic receptor (OORTGIESEN, 1993). When studying the function of the basal activity of erythrocyte Ca/ATPase in Nigerian workers occupationally exposed to lead, studies indicate that there was a reduction in the pump's function correlated with blood lead levels. This result indicates a possible competition between lead and calcium for the pump (IMAH-HARRY, 2012). In addition, lead may serve as a substitute for calcium in calcium-dependent signaling pathways and interact with calmodulin, PKC and calcium-dependent potassium channels (SIMONS, 1991; HABERMANN,

1983; AND RICHARDT, 1986). Thus, the interactions of lead with cellular Ca^2 + through these complex mechanisms in vascular cells may contribute to alterations in vascular resistance.

Dopamine is known to be a potent renal vasodilator, increasing GFR and RBF. Dopamine acts via the D1 receptor, promoting renal vasodilation and natriuresis. It is also capable of inhibiting the proliferation of mesangial cells (MASAYO, 2012). Studies have shown that in rats treated orally with low doses of lead acetate (0.1%), their offspring showed an increase in D2 receptors (LISTOS et al., 2013). Another study shows that simultaneous and early exposure to low levels of lead and other metals causes an increase in the anxiety behavior of adult rats analyzed in the elevated cross maze, in conjunction with an increase in dopamine and serotonin levels (LERET et al., 2003).

Controversially, recent experimental studies have shown that the initial mechanism of the alteration in glomerular hemodynamics may well depend on the production of reactive oxygen species, induced by exposure to lead, with *upregulation* of cyclooxygenase (COX-2) expression in the vascular smooth muscle wall (COURTOIS ET AL, 2003). Studies have also shown an increase in lead-induced thromboxane production in workers exposed to this metal (ROELS ET AL., 1994). Cardenas and Hotter found a significant increase in the urinary excretion of vasoconstrictor prostaglandin metabolites and thromboxane (TXB2), when compared to the values found in control workers. In addition, Dormam and Freeman demonstrated that lead promotes the *in vitro* release of arachidonic acid by smooth muscle cells *in the* vasculature, through the activation of phospholipase A2. Khalil-Manesh studied the effect of exposure to small and large concentrations of lead (100 vs. 5000 ppm in water over a period of 1-12 months) on plasma endothelin (ET) in rats. Rats exposed to low levels of lead showed a significant increase in plasma ET-3 concentration. Similarly, Gonick et al (1997) showed a significant increase in plasma and urine ET-3 in rats with lead-induced hypertension. In a recent study Molero et al (2006) provided evidence that lead exposure can increase endothelin activity in the vascular tissue of rats.

When we evaluated the filtration fraction of the animals treated with lead, we observed that there was no difference between the groups. As the filtration fraction is calculated as a function of renal plasma flow and glomerular filtration rate, and both parameters increased, there was no change in this parameter.

In our study, we observed a reduction in blood pressure in rats after lead acetate injection. This result corroborates the data on a reduction in renal vascular resistance, and we can speculate that a reduction in peripheral vascular resistance is occurring, contributing to the drop in blood pressure.

In contrast to our data, Fiorim (2011) showed that treatment with lead to maintain its blood concentration at around 9.98 mg/dl results in an increase in systolic blood pressure. Another study showed that acute administration of lead in low concentrations increased left and right ventricular systolic pressure and produced a positive inotropic and lusitropic effect on the heart muscle, both in *vivo* and *in vitro* (FIORESI, 2013).

Several clinical and animal studies have focused on the sympathetic system as a possible mediator of lead-induced hypertension. In a group of workers exposed to lead, Chan (1996) found increases in the plasma concentration of norepinephrine. Similarly, Tsao et al in 2000 demonstrated that plasma norepinephrine was elevated in lead-treated rats, which also exhibited significant reductions in β-adrenergic receptor density and isoproterenol-stimulated cAMP production in the heart and aorta. The ability to rapidly stimulate sympathetic nervous system activity was confirmed by Lai (2002), through intra-thecal (IT) injection of PbCl2 in rats. Together, these results provide evidence for the stimulating effect of lead on the sympathetic nervous system and its contribution to the cardiovascular effects of lead exposure.

We can't rule out the possibility that lead may cause changes in the volume of extracellular fluid. Corroborating this hypothesis, Giridhar and Isom studied rats treated with an IP injection of lead acetate at a dose of 0.0-1.0 mg/Kg twice a week for 30 days. The lead-exposed animals exhibited fluid retention and a dose-dependent decrease in plasma ANP concentration. They attributed lead toxicity as a crucial factor in the dysregulation of Atrial Natriuretic Peptide (ANP) production. ANP is a vasodilator and a natriuretic agent, which is produced and secreted by cardiac myocytes.

Changes in vascular reactivity induced by lead may be related to changes in blood pressure and renal hemodynamics. The rapid action of lead on vascular reactivity varies according to the concentration of lead used and the species of animal studied (VAZIRI&GONICK, 2008). It has been shown that the addition of lead acetate causes a metal concentration-dependent vasoconstriction in isolated rabbit mesenteric artery through PKC activation (WATTS, 1995). Similarly, a vasoconstrictor response dependent on the concentration of lead acetate (0.1-3.1 mM) was observed in Wistar rat thoracic aortic rings when intact or when the endothelium was removed (VALENCIA, 2001). In contrast to these results, other studies conclude that lead acetate does not cause contraction of rat aortic rings, as reported in the study by Shelkovnikov and Gonick (2001). Furthermore, lead acetate did not modify the response to norepinephrine, phorbol ester or isoproterenol, but did increase the contractile response to sub-maximal concentrations of calcium.

In an effort to discern the possible effect of lead exposure on vascular reactivity, *Purdy et all* compared the responses to different vasoactive agents in Sprague-Dawley rats with lead-induced hypertension. They found no significant difference in the vasoconstrictor response to noradrenaline and phenylephrine or vasodilator response to acetylcholine and nitroprusside in the aortic rings. In contrast, Marques et al showed a significant reduction in the vasodilatory response to acetylcholine and nitroprusside in rats with lead-induced hypertension. It should be noted that the Wistar rats used in the previous study had been given 5 ppm of lead acetate in their drinking water for a month, while the rats used in the study by Purdy et al had been given a higher dosage (100 ppm) for a longer period (3 months). Therefore, the concentration and duration of exposure may explain the differences observed between the two studies. In addition, the effect of lead on vascular reactivity can vary from one tissue to another. This is exemplified by the studies of Oishi et all with rats exposed to tetra-ethyl lead for 3 months. They found a significant reduction in acetylcholine-mediated vessel relaxation in the presence of the NOS inhibitor in the mesenteric artery, but not in the aorta of the same animals. These observations suggest that lead exposure may damage the endothelium in the rat mesenteric artery, but not the aorta.

Endothelial damage or dysfunction can result in atherosclerosis, thrombosis and tissue damage. Chronic exposure to lead has been shown to promote atherosclerosis in experimental animals (REVIS, 1981). Using cultured endothelial cells, Kaji et al showed that lead nitrate causes endothelial monolayer denudation, reduces proliferation and attenuates basic fibroblast growth factor (bFGF) and acidic fibroblast growth factor (aFGF) in bovine aortic endothelial cell culture. In addition, other studies have shown that lead acetate (1-100 IM) causes a time- and concentration-dependent inhibition of angiogenesis (FUJIWARA, 1998; UEDA,1997; KISHIMOTO, 1997) and interferes with the synthesis of glycosaminoglycans and various in endothelial cell cultures (FUJIWARA1999; KAJI, 1991; FIJIWARA, 1995). Given the crucial role of proteoglycans in many aspects of endothelial and vascular function, the impairment of their production by lead plays an important role in the adverse actions of lead exposure. Lead stimulates the proliferation of vascular smooth muscle cells (VSMC) in the bovine aorta, and this increase was induced by an increase in bFGF (FIJIWARA, 1995).

Lead can also alter the shape of VSMC cells, promoting their transformation from a spindle shape to a parallelepiped shape, which simulates the morphology of a neo-intimal cell (CARSIA, 1995). In addition, Yamamoto et al found that lead causes a significant inhibition of tissue plasminogen activator (t-PA). Thus, exposure to lead appears to evoke a negative effect on the fibrinolytic process in the cellular constituents of sub-endothelial tissue.

In our study, we also observed a reduction in uremia in the animals that received this metal. Corroborating our data, Patrick (2006) also observed this decrease in plasma urea levels in animals treated with lead. This result was associated with a damaging action of lead on the cells of the proximal convoluted tubule, which decreases its reabsorption and increases its elimination, thus reducing its concentration in the blood. Lead can cause inflammation in the cells of the proximal tubule via oxidative stress (VAZIRI, 2006), which could reduce the reabsorption of urea, contributing to its lower plasma concentration found in our study. As is already known, lead can cause oxidative stress in various animal tissues. Oxidative stress activates nuclear factor kappa B (NFkB), which is the general transcription factor for numerous cytokines, chemokines and adhesion molecules. Thus, through the activation of NFkB, oxidative stress can cause inflammation. On the other hand, inflammation can cause oxidative stress via the generation and release of ROS by activated immune cells. It is therefore intuitive that, by promoting the production of ROS, lead can provoke a perpetual cycle of oxidative stress and inflammation in the target tissues. Studies confirm (RODRIGUEZ-ITURBE, 2004) the activation of NFkB in the renal tubules, infiltration of T-cells, macrophages and angiotensin II-positive cells, an increase in apoptotic cells and accumulation of nitrotyrosine in the kidneys of rats treated with lead.

6. CONCLUSION

In conclusion, our data show that acute administration of a low dose of lead acetate can cause deleterious effects on kidney function. These effects are reflected in increased glomerular filtration rate, measured by increased inulin *clearance* and reduced uremia. Increased renal perfusion is demonstrated by increased renal plasma flow and renal blood flow. Treatment indicates action on the renal vascular bed with a reduction in renal vascular resistance. These data demonstrate the toxic potential of lead on renal hemodynamics, even in the case of acute, low-dose exposure.

In particular, renal hyperfiltration appears to be a rapid response to release lead from the body, since the metal is associated with subtle tubular effects. These results with low levels of environmental exposure, sometimes with no detectable limit, reinforce the need to control and regulate potential sources of heavy metal contamination.

7. BIBLIOGRAPHICAL REFERENCES

Alves GM, Barao MA, Odo LN, Nascimento Gomes G, Franco Md Mdo C, Nigro D, Lucas SR, Laurindo FR, Brandizzi LI, Zaladek Gil F (2002). *L-Arginine effects on blood pressure and renal function of intrauterine restricted rats.* Pediatr Nephrol, 17:856-862

ATSDR. Toxicological profile for Lead. US Department of Health and Human Services. Public Health Service. Agency for Toxic Substances and Disease Registry, 1993.

Baird C. Toxic heavy metals. IN. Quimica Ambiental, 2ed. Sao Paulo: Bookman, 2002. p 403-439.

Boscolo P, Carmignani M. Neurohumoral blood pressure regulation in lead exposure. Environ Health Perspect 1988; 78: 101-106, doi: 10.1289/ehp.8878101.

Cardenas A, Roels H, Bernard AM, Barbon R, Buchet JP, Lauwerys RR, *et al.* Markers of early renal changes induced by industrial pollutants. II. Application to workers exposed to lead. *Br J Ind Med* 1993; *50* : 28-36.

Carsia RV, Forman D, Hock CE, Nagele RG, McIlroy PJ. Lead alters growth and reduces angiotensin II receptor density of rat aortic smooth muscle cells. *Proc Soc Exp Biol Med* 1995; *210* : 180-90.

Chang HR, Chen SS, Chen TJ, Ho CK, Chiang HC, Yu HS. Lymphocyte β2-adrenergic receptors and plasma catecholamine levels in lead-exposed workers. *Toxicol Appl Pharmacol* 1996; *139* : 1-5.

Chang HR, Chen SS, Tsao DA, Cheng JT, Ho CK, Yu HS. Change of cardiac β-adrenoceptors in lead-exposed rats. *Toxicology* 1997; *123* : 27-32.

Cory-Slechta DA, Pokora MJ, and Johnson JL (1996) Postweaning lead exposure enhances the stimulus properties of N-methyl-aspartate: possible dopaminergic involvement? Neurotoxicology 17(2): 509-521.

Courtois E, Marques M, Barrientos A, Casado S, Lopez-Farre A. 2003. Lead-induced downregulation of soluble guanylate cyclase in isolated rat aortic segments mediated by reactive oxygen species and cyclooxygenase-2. J Am Soc Nephrol 14:14641470.

Dorman RV, Freeman EJ. Lead-dependent effects on arachidonic acid accumulation and the proliferation of vascular smooth muscle. *J Biochem Mol Toxicol* 2002; *16* : 245-53.

Eaton, DC; Pooler, JP. Vander's Renal Physiology. 6.ed - Porto Alegre: Artmed, 2006.

Farmand F, Ehdaie A, Roberts CK, Sindhu RK. Leadinduced dysregulation of superoxide dismutases, catalase, glutathione peroxidase, and guanylate cyclase. Environ Res 2005; 98: 33-39, doi: 10.1016/j.envres.2004.05.016.

Fernandez R, Malnic G (1995). *Secretion of HCO3-/OH- in cortical distal tubule of the rat.* J Memb Biol, 143: 227-235

Fiorim J, Ribeiro Ju' nior RF, Silveira EA, Padilha AS, Vescovi MVA, et al. (2011) Low-Level Lead Exposure Increases Systolic Arterial Pressure and Endothelium-Derived Vasodilator Factors in Rat Aortas. PLoS ONE 6(2): e17117. doi:10.1371/journal.pone.0017117

Fiorim J, Ribeiro Ju' nior RF, Silveira EA, Padilha AS, Vescovi MVA, et al. (2011) Low-Level Lead Exposure Increases Systolic Arterial Pressure and Endothelium-

Derived Vasodilator Factors in Rat Aortas. PLoS ONE 6(2): e17117. doi:10.1371/journal.pone.0017117

Friedman AN. High-protein diets: potential effects on the kidney in renal health and disease. 2004. Am J Kidney Dis 44:950-962.

Fuehr J, Kaczmarczyk J, Kruttgen CD (1955). *A simple colorimentric method of inulin in renal clearance studies on metabolically normal subjects and diabetics.* Wien Klin Wochenschr, 33: 729-730

Fujiwara Y, Kaji T, Yamamoto C, Sakamoto M, Kozuka H. Stimulatory effect of lead on the proliferation of cultured vascular smooth-muscle cells. *Toxicology* 1995; *98* : 105-10.

Fujiwara Y, Kaji T, Yamamoto C, Sakamoto M, Kozuka H. Stimulatory effect of lead on the proliferation of cultured vascular smooth-muscle cells. *Toxicology* 1995; *98* : 105-10.

Fujiwara Y, Kaji T. Possible mechanism for lead inhibition of vascular endothelial cell proliferation: a lower response to basic fibroblast growth factor through inhibition of heparan sulfate synthesis. *Toxicology* 1999; *133* : 147-57.

FujiwaraY, Watanabe S, Sakamoto M, Kaji T. Repair of wounded monolayers of cultured vascular endothelial cells after simultaneous exposure to lead and zinc. *Toxicol Lett* 1998; *94* : 181-8.

Giridhar J, Isom GE. Interaction of lead acetate with atrial natriuretic factor in rats. *Life Sci* 1990; *46* : 569-76.

Goldstein GW (1993) Evidence that lead acts as a calcium substitute in second messenger metabolism. Neurotoxicology 14(2-3): 97-101.

Goldstein GW. Evidence that lead acts as a calcium substitute in second messenger metabolism. Presented at: Ninth International Neurotoxicology Conference; October 1991; Little Rock, AR. *Neurotoxicology* 1993; *14* : 97-101.

Gonick HC, Ding Y, Bondy SC, Ni Z, Vaziri ND. Leadinduced hypertension: interplay of nitric oxide and reactive oxygen species. Hypertension 1997; 30: 1487-1492, doi: 10.1161/01.HYP.30.6.1487.

Gonick HC. Ding Y, Bondy SC, Ni Z, Vaziri ND. Lead-induced hypertension: interplay of nitric oxide and reactive oxygen species. *Hypertension* 1997; *30* : 1487-92.

Grizzo LT, Cordellini S. Perinatal lead exposure affects nitric oxide and cyclooxygenase pathways in aorta of weaned rats. Toxicol Sci 2008; 103: 207-214, doi: 10.1093/toxsci/ kfn018.

Gruden N, Buben M (1977) Influence of lead on calcium metabolism. Bulletin of Environmental Contamination and Toxicology 18(3): 303-307.

Habermann E, Crowell K, Janicki P. Lead and other metals can substitute for Ca2+ in calmodulin. *Arch Toxicol* 1983; *54* : 61-70.

Hotter G, Fels LM, Closa D, Roselló J, Stolte H, Gelpi E. Altered levels of urinary prostanoids in lead-exposed workers. *Toxicol Lett* 1995; *77* : 309-12.

Imah-Harry JU, Akiibinu MO, Omotosho IO, Olorunsogo OO. (2012) Human erythrocyte membrane Ca(2+)-ATPase during occupational exposure to lead. Afr J Med Med Sci. 2012 Dec;41 Suppl:91-7.

Kaji T, Fujiwara Y, Hoshino M, Yamamoto C, Sakamoto M, Kozuka H. Inhibitory effect of lead on the proliferation of cultured vascular endothelial cells. *Toxicology*

1995; *95* : 87-92.

Kaji T, Suzuki M, Yamamoto C, Mishima A, Sakamoto M, Kozuka H. Severe damage of cultured vascular endothelial cell monolayer after simultaneous exposure to cadmium and lead. *Arch Environ Contam Toxicol* 1995; *28* : 168-72.

Kaji T, Yamamoto C, Sakamoto M. Effect of lead on the glycosaminoglycans metabolism of bovine aortic endothelial cells in culture. *Toxicology* 1991; *68* : 249-57.

Khalil-Manesh F, Gonick HC, Weiler EW, Prins B, Weber MA, Purdy RE. Lead-induced HTN: possible role of endothelial factors. *Am J Hypertens* 1993; *6* : 723-9.

Khalil-Manesh F, Gonick HC, Weiler EW, Prins B, Weber MA, Purdy RE. Lead-induced hypertension: possible role of endothelial factors. Am J Hypertens 1993; 6: 723-729.

Kishimoto T, Oguri T, Ueda D, Tada M. Effect of lead on tube formation by cultured human vascular endothelial cells. *Arch Toxicol* 1995; *69* : 718-21.

Lai CC, Lin HH, Chen CW, Chen SH, Chiu TH. Excitatory action of lead on rat sympathetic preganglionic neurons *in vitro* and *in vivo. Life Sci* 2002; *71* : 1035-45.

Leret, M. L.; San Millàn, J. A.; Antonio, T. Perinatal exposure to lead and cadmium affects anxiety-like behavior. Toxicology 186: 125-130, 2003.

Listos J, Baranowska-Bosiacka I, Talarek S, Listos P, Orzelska J, Fidecka S, Gutowska I, Kolasa A, Rybicka M, Chlubek D. The effect of perinatal lead exposure on dopamine receptor D2 expression in morphine dependent rats. Toxicology. 2013 Aug 9;310:73-83. doi: 10.1016/j.tox.2013.05.007. Epub 2013 May 20.

Lustberg M, Silbergeld E (2002). Blood lead levels and mortality. Arch Intern Med;

162: 2443-2449, doi: 10.1001/archinte.162.21.2443.

Lustberg M, Silbergeld E (2002). Blood lead levels and mortality. Arch Intern Med; 162: 2443-2449, doi: 10.1001/archinte.162.21.2443.

Lyn Patrick, ND Lead Toxicity, A Review of the Literature. Part I: Exposure, Evaluation, and TreatmentAlternative Medicine Review u Volume 11, Number 1 u 2006

M. Fioresi; L.B. Furieri; M.R. Simóes; R.F. Ribeiro Junior; E.F. Meira; A.A. Fernandes; I. Stefanon; D.V. Vassallo(2013). *Acute exposure to lead increases myocardial contractility independent of hypertension development.* Brazilian Journal of Medical and Biological Research, 00: 1-8, http://dx.doi.org/10.1590/1414-431X20122190 ISSN 1414-431X

Marchioro M, Swanson KL, Aracava Y, and Albuquerque EX (1996) Glycine and calcium-dependent effects of lead on N-methyl-aspartate receptor function in rat hippocampal neurons. Journal of Pharmacol and Experimental Therapeutics 279(1): 143-153.

Marques M, Millas I, Jiménez A, Garcia-Colis E, Rodriguez- Feo JA, Velasco S, *et al.* Alteration of the soluble guanylate cyclase system in the vascular wall of lead - induced hypertension in rats. *J Am Soc Nephrol* 2001; *12* : 2594-600.

Marques M, Millas I, Jiménez A, Garcia-Colis E, Rodriguez-Feo JA, Velasco S, *et al.* (2001) *Alteration of the soluble guanylate cyclase system in the vascular wall of lead - induced hypertension in rats. J Am Soc Nephrol*; *12* : 2594-600.

Masayo Hirose, Fumihiro Tomoda, Tsutomu Koike, Hidenori Yamazaki, Maiko Ohara, Hexing Liu,Satoshi Kagitani, Hiroshi Inoue. *Imbalance of Renal Production*

Between5-Hydroxytryptamine and Dopamine in Patients With Essential Hypertension *Complicated by Microalbuminuria* Am J Hypertens. 2013 Feb;26(2):227-33. doi: 10.1093/ajh/hps008. Epub 2012 Dec 28.

Midio, AF; Martins, DI.; Toxic agents directly contaminating food. In: Food Toxicology. Sao Paulo, Varela, 2000. Chap 3 p 61-130.

Molero L, Carrasco C, Marques M, Vaziri ND, Mateos-Caceres PJ, Casado S, *et al.* Involvement of endothelium and endothelin-1 in lead-induced smooth muscle cell dysfunction in rats. *Kidney Int* 2006; *69* : 685-90.

Navas-Acien A, Guallar E, Silbergeld EK, Rothenberg SJ. Lead exposure and cardiovascular disease - A systematic review. *Environ Health Perspect* 2007; *115* : 472-82.

Oishi H, Nakashima M, Totoki T, Tomokuni K. Chronic lead exposure may inhibit endothelium dependent hyperpolarizing factor in rats. *J Cardiovasc Pharmacol* 1996; *28* : 558-63.

Oortgiesen M, Leinders T, van Kleef RG, and Vijverberg HP (1993) Differential neurotoxicological effects of lead on voltage-dependent and receptor-operated ion channels. Neurotoxicology 14(2-3): 87-96.

PAOLIELLO, M. M. B.; CHASIN, A. A. M. Ecotoxicologia do chumbo e seus compostos. Salvador: CRA, 2001. 144 p.

PATOCKA, J., CERNY, K. Inorganic lead toxicology. Acta Medica (Hradec Kralove). v. 46, n.2, p.65-72, 2003.

Patrick L. (2006) *Lead Toxicity, a review of the literature. Part I: Exposure, Evaluation, and treatment.* Alternative Medicine Review 11: 2-22.

Patrick L. (2006) *Lead Toxicity, a review of the literature. Part I: Exposure, Evaluation, and treatment.* Alternative Medicine Review 11: 2-22.

Patrick L. (2006) *Lead Toxicity, a review of the literature. Part I: Exposure, Evaluation, and treatment.* Alternative Medicine Review 11: 2-22.

Purdy RE, Smith JR, Ding Y, Oveisi F, Varizi ND, Gonick HC. Lead-induced hypertension is not associated with altered vascular reactivity *in vitro*. *Am J Hypertens* 1997; *10* : 997-1003.

Revis NW, Zinsmeister AR, Bull R. Atherosclerosis and hypertension induction by lead and cadmium ions: an effect prevented by calcium ion. *Proc Natl Acad Sci USA* 1981; *78* : 6494-8.

Richardt G, Federolf G, Habermann E. Affinity of heavy metal ions to intracellular Ca2+-binding proteins. *Biochem Pharmacol* 1986; *35* : 1331-5.

Rodriguez-Iturbe B, Sindhu RK, Quiroz Y, Vaziri ND. Chronic exposure to low doses of lead results in renal infiltration of immune cells, NF-êB activation, and overexpression of tubulointerstitial angiotensin II. *Antioxid Redox Signal* 2005; *7* : 1269-74.

Rodriguez-Iturbe, B, Vaziri ND, Herrera-Acosta J, Johnson RJ. Oxidative stress, renal infiltration of immune cells, and salt-sensitive hypertension: all for one and one for all. *Am J Physiol* 2004; *286* : F606-16.

Roels H, Lauwerys R, Konings J, Buchet JP, Bernard A, Green S, et al. 1994. Renal function and hyperfiltration capacity in lead smelter workers with high bone lead. Occup Environ Med 51:505-512.

Saryan LA, Zensz C. Lead and its compounds, pp. 506-541, In C Zensz, OB

Dickerson & EP Horvath (eds.) Occupational medicine. Ed Mosby-Year Book, USA, 1994.

Shelkovnikov SA, Gonick HC(2001). Influence of lead on rat thoracic aorta contraction and relaxation. *Am J Hypertens*; *14* :873-8.

Simoes MR, Ribeiro Junior RF, Vescovi MV, de Jesus HC, Padilha AS, Stefanon I, et al. Acute lead exposure increases arterial pressure: role of the renin-angiotensin system. PloS One 2011; 6: e18730, doi: 10.1371/journal.pone.0018730.

Simoes MR, Ribeiro Júnior RF, Vescovi MVA, de Jesus HC, Padilha AS, et al. (2011) *Acute Lead Exposure Increases Arterial Pressure: Role of the Renin-Angiotensin System.* PLoS ONE 6(4): e18730. doi:10.1371/journal.pone.0018730.

Simoes MR, Ribeiro Júnior RF, Vescovi MVA, de Jesus HC, Padilha AS, et al. (2011) *Acute Lead Exposure Increases Arterial Pressure: Role of the Renin-Angiotensin System.* PLoS ONE 6(4): e18730. doi:10.1371/journal.pone.0018730.

Simons TJB. Lead transport and binding by human erythrocytes *in vitro*. *Pflugers Arch* 1993; *423* : 307-13.

Simons TJB. Lead-calcium interactions in cellular lead toxicity. Presented at: Ninth international neurotoxicology conference; October 1991; Little Rock, AR. *Neurotoxicology* 1996; *14* : 77-86.

Smith HW, Finkelstein N, Aliminosa L, Crawford B, Graber M (1945). *The renal clearances of substituted hippuric acid derivatives and other aromatic acids in dog and man.* J Clin Invest, 24(3): 388-404

SPiNOLA, A. G.; FERNiCOLA, N.A. G. G., MENDES, R. 1980. Occupational lead poisoning. In: Occupational Medicine - Occupational Diseases. Sao Paulo: Sarvier.

p. 437-460.

Tsao DA, Yu HS, Cheng JT, Ho CK, Chang HR. The change of beta-adrenergic system in lead-induced hypertension. Toxicol Appl Pharmacol 2000; 164: 127-133, doi: 10.1006/ taap.1999.8871.

Tsao DA, Yu HS, Cheng JT, Ho CK, Chang HR. The change of β-adrenergic system in leadinduced hypertension. *Toxicol Appl Pharmacol* 2000; *164* : 127-33.

Ueda D, Kishimoto T, Dekio S, Tada M. Inhibitory effect of lead on tube formation by cultured human vascular endothelial cells. *Hum Cell* 1997; *10* : 283-91.

Valencia I, Castillo EE, Chamorro G, Bobadilla RA, Castillo C. (2001) Lead induces endothelium- and Ca2+-independent contraction in rat aortic rings. *Pharmacol Toxicol; 89* :177-82.

Vaziri ND, Ding Y, Ni Z, Gonick HC. Altered nitric oxide metabolism and increased oxygen free radical activity in lead-induced hypertension: effect of lazaroid therapy. Kidney Int 1997; 52: 1042-1046, doi: 10.1038/ki.1997.426.

Vaziri ND, Ding Y, Ni Z. Compensatory up-regulation of nitric-oxide synthase isoforms in lead-induced hypertension; reversal by a superoxide dismutase-mimetic drug. J Pharmacol Exp Ther 2001; 298: 679-685.

Vaziri ND, Rodriguez-Itrube B. Mechanisms of disease: oxidative stress and inflammation in the pahtogenesis of hypertension. *Nat Clin Pract Nephrol* 2006; *2* : 582-93.

Vaziri, ND; Gonick, HC (2008). Cardiovascular effects of the lead exposure. Indian J Med Res 128, October, pp 426-435.

Watts SW, Chai S, Webb RC (1995). Lead acetate-induced contraction in rabbit mesenteric artery: interaction with calcium and protein kinase C. *Toxicology; 99* : 5565.

Weaver VM, Lee BK, Ahn KD, Lee GS, Todd AC, Stewart WF, et al. 2003a. Associations of lead biomarkers with renal function in Korean lead workers. Occup Environ Med 60:551-562.

Weaver VM, Lee BK, Ahn KD, Lee GS, Todd AC, Stewart WF, et al. 2003a. Associations of lead biomarkers with renal function in Korean lead workers. Occup Environ Med 60:551-562.

Weiler E, Khalil-Manesh F, Gonick HC. Effects of lead and a low-molecular-weight endogenous plasma inhibitor on the kinetics of sodium-potassium-activated adenosine triphosphatase and potassium-activated p-nitrophenylphosphatase. Clin Sci 1990; 79: 185-192.

WHO, World Health Organization. Environmental Health Criteria - 165 Inorganic Lead. Geneva: WHO, 1995

Xie Y, Chiba M, Shinohara A, Watanabe H, Inaba Y, Studies on lead-binding protein and interaction between lead and selenium inde the human erythrocytes. Industrial Health 36:234-239, 1998.

Yamamoto C, Miyamoto A, Sakamoto M, Kaji T, Kozuka H. Lead perturbs the regulation of spontaneous release of tissue plasminogen activator and plasminogen activator inhibitor-1 from vascular smooth muscle cells and fibroblasts in culture. *Toxicology* 1997; *117* : 153-61.

MIX
Papier aus verantwortungsvollen Quellen
Paper from responsible sources
FSC® C105338

Printed by Books on Demand GmbH, Norderstedt / Germany